Tracey Rosenlicht

How Caffeine Effects Heart Rate

GRIN Verlag

Bibliografische Information der Deutschen Nationalbibliothek:

Die Deutsche Bibliothek verzeichnet diese Publikation in der Deutschen National-
bibliografie; detaillierte bibliografische Daten sind im Internet über http://dnb.d-
nb.de/ abrufbar.

Imprint:

Copyright © 2012 GRIN Verlag GmbH
Druck und Bindung: Books on Demand GmbH, Norderstedt Germany
ISBN: 978-3-656-36182-4

This book at GRIN:

http://www.grin.com/en/e-book/207849/how-caffeine-effects-heart-rate

GRIN - Your knowledge has value

Der GRIN Verlag publiziert seit 1998 wissenschaftliche Arbeiten von Studenten, Hochschullehrern und anderen Akademikern als eBook und gedrucktes Buch. Die Verlagswebsite www.grin.com ist die ideale Plattform zur Veröffentlichung von Hausarbeiten, Abschlussarbeiten, wissenschaftlichen Aufsätzen, Dissertationen und Fachbüchern.

Visit us on the internet:

http://www.grin.com/

http://www.facebook.com/grincom

http://www.twitter.com/grin_com

Tracey Rosenlicht

IB Bio 1
5 January 2013

DESIGN

ASPECT 1: DEFINING THE PROBLEM AND SELECTING VARIABLES

Background Information: The human heart is a major muscular organ located in the thoracic
cavity between the lungs. Its major function is to pump blood throughout the body. A double-
layered sac, Pericardium, which is the tough connective tissue protects and anchors the heart.
There is fluid between the layer of the sac allow for lubrication of the heart's continual motions.
The inner layer of the sac is the heart wall, which is mainly cardiac muscle. The human heart is
made up of two chambers. The atriums receive blood from veins, and the ventricles pump blood
into the arteries. For the blood to pass through the an atrium to a ventricle for example, the blood
has to pass through a heart valve. Valves control the blood from moving backwards. The "lub-
dub" sound made by a beating heart derives from the closing of the atrioventricular (AV) valves,
then the concurrent closing of the aortic and pulmonary valves (Starr, 2007). The human heart is
also myogenic meaning the heart is independent of an outside stimulus from the nervous system.
The sinoatrial (SA) node, pace maker, is responsible for sending electrical impulses through the
heart making it contract and pump blood. The human heart is very much affected by the
consumption of caffeine. Caffeine can be found in certain coffees, teas, sodas, and chocolates.
By consuming caffeine one's heart rate to dramatically increase and also cause abnormal heart
rhythms (*Medline Plus: Caffeine*, 2012).

Problem Question: What is the effect of caffeine from coffee, herbal tea, and black tea on heart
rate?

Hypothesis: If a human consumes a beverage containing a higher concentration of caffeine, then
the heart rate will be higher than consuming a beverage with a lower concentration of caffeine.

Hypothesis Explanation: This is because caffeine is a stimulant. Stimulants change the way the
brain works by changing the way nerve cells communicate. Nerve cells, called neurons, send
messages to each other by releasing chemicals called neurotransmitters. Neurotransmitters work
by attaching to key sites on neurons called receptors. Stimulants can also cause the body's blood
vessels to narrow, constricting the flow of blood, which forces the heart to work harder to pump
blood through the body. The heart may work so hard that it temporarily loses its natural rhythm
(*Stimulants*, 2012). A normal resting heart rate for adult's ranges from 60 to 100 beats a minute.
In most cases, a lower heart rate at rest implies more efficient heart function and better
cardiovascular fitness. For example, a well-trained athlete might have a normal resting heart rate
closer to 40 beats a minute (*Caffeine content for coffee, tea, soda and more*, 2011). Additionally
caffeine blocks the enzyme phosphodiesterace, which normally activates production of an
enzyme, cyclic AMP (cAMP), this initates a protein that increases heart rate. The heart rate is
regulated as cAMP is kept in a drug free cell. When caffeine is introduced, cAMP is eliminated,

placing the protein (PKA) into overdrive. Therefore, heart rate is increased when caffeine is consumed (*Caffeine Pharmacology*, 2012). Nonetheless a beverage with more caffeine will lead to a greater amount of heart rate in beats per minute (bpm±1.0bpm) as compared to a beverage with less caffeine.

Variables
      Dependent: Number of heart pulse rate in beats per minute (bpm±1.0bpm). This measurement will be taken by placing the index and middle finger on the neck pressing gently on the carotid artery, ten minutes after the consumption of either caffeinated coffee, herbal tea, or black tea, and counting the heart rate over the course of one minute.
      Independent: Amount of caffeine found in 250ml of either caffeinated coffee, herbal tea or black tea.

ASPECT 2: CONTROLLING VARIABLES

Table 1. Controlled variables kept constant

CONTROLLED VARIABLES	WHY it must be controlled	HOW it is controlled
Amount of time after caffeine is ingested to record number of heart pulse rate in beats per minute (bpm±1.0bpm).	Amount of time after caffeine is ingested to record number of heart pulse rate in beats per minute (bpm±1.0bpm), must be controlled because caffeine affects heart rate ten minutes after it is ingested (Adan 2008). Yes, every person metabolizes caffeine differently, although this allows for a more controlled experiment.	This variable is controlled by using a stopwatch to record a ten-minute time span immediately after the subject finishes the 150ml beverage.
Brewing time of black and herbal tea, recorded in minutes.	The brewing time of black and herbal tea, recorded in minutes, must be controlled. This is because based on the chart below, 5 minutes produces the greatest caffeine (mg) content (Chin 2008).	This variable is controlled by using a stopwatch to record the ten-minute time span of the teabag submerged in the boiled water (100° C). The temperature will be measured using a cooking thermometer.
Amount of caffeinated coffee, black tea, and herbal tea consumed (150ml).	The amount of every beverage consumed (150ml) must be controlled because although black tea has more caffeine than herbal tea, if at any point one of the subjects consume a greater volume of herbal tea over the black tea, the data will be invalid.	This variable is controlled by using a liquid measuring cup in milliliters.

			(Pyrex 2013).

Experimental Control: The experimental control is the number of heart pulse rate in beats per minute (bpm±1.0bpm) after an hour of physical rest. The experimental variables do not effect the experimental control. The experimental control is to be used in comparison against other trials with physical activity.

ASPECT 3: DEVELOPING A METHOD FOR COLLECTION OF DATA

Apparatus and Materials: The materials necessary for this experiment include a metric liquid measuring cup, Twinings black tea, Twinings Pure Chamomile tea, Folgers Classic Roast Caffeinated Coffee, five stopwatches, hot water, cooking gloves, tea kettle, a graph to record data, and subjects to participate in experiment,.

Figure 1. Picture depicting neck pulse taking technique.

Figure 2. Picture of Twininings black and chamomile tea, and Folgers "Classic Roast" caffeinated coffee.

Procedure:
Note: All data and/or observations should be noted on a data table.

Safety Precaution: Be careful when using minors as subjects, for caffeine is a stimulant and can cause cardiac arrhythmia. Always receive parental consent on all minors. Consider any food allergies.

1. Have five subjects sitting at a table (or however many desired). With a stopwatch placed next to each person
2. Have subject take their pulse after having rested for one hour. Take heart rate by placing the index finger and middle finger on the outside of the neck, where the carotid artery is located, in beats per minute (bpm±1.0bpm). Record the data immediately by writing down the number of heart beats per second (bpm±1.0bpm) on a piece of paper.
3. Boil a teakettle containing around 800ml of water, so that if any amount of water evaporates there should still be at least 750ml of water left. To account for the 150ml of water to be drank by each person (150ml*5=750). Take temperature of water with cooking thermometer, it should read around 55°C.
4. Once the teakettle has boiled, place one mug in front of each person with the Twinings Black Teabag inside. Carefully measure 150ml of hot water in a Pyrex glass-measuring cup, while wearing cooking gloves. In order to not burn oneself.
5. Once all five cups are filled with 150 ml of hot water, use the stopwatch to time three minutes.
6. Once the stopwatch reaches three minutes take out the teabag, so that the caffeine content is not affected by brewing time.
7. Once the subject determines that the tea is at a reasonable temperature, they may drink the tea.
8. Since every person has their own stopwatch, they can drink the tea at their own pace. They will start the stopwatch as soon as they finish drinking their tea. Once they reach ten minutes they will take their heart rate by placing their index finger and middle finger on the neck outside of the carotid artery for twenty seconds and then multiply that number by three to get the number of heart beats per second (bpm±1.0bpm). Record the data immediately by writing the number of heart beats per second (bpm±1.0bpm) down.
9. Repeat steps 1-8 to determine the effects of the Twinings Herbal Tea. Although it is best to run this trial on a separate day because everyone metabolizes caffeine differently.
10. Repeat steps 1-2 to determine the effects of the Folgers Caffeinated Coffee. Although it is best to run this trial on a separate day because everyone metabolizes caffeine differently.
11. Brew 150ml of coffee per person (750ml in total).
12. Place one mug in front of each person. Carefully measure 150ml (per person) of hot water in a Pyrex glass-measuring cup, while wearing cooking gloves. In order to not burn oneself. Take temperature of water with cooking thermometer, it should read around 55°C.
13. Have the subject start their stopwatches immediately after consuming the entire drink.
14. Have subject take their pulse ten minutes after they have finished their coffee. Take heart rate by placing the index finger and middle finger on the outside of the neck, where the carotid artery is located, in beats per minute (bpm±1.0bpm). Record the data immediately, by writing the number of heart beats per second (bpm±1.0bpm) down on paper.
15. Repeat steps 1-14 once more to accumulate twice the amount of data.

DATA COLLECTION AND PROCESSING
ASPECT 1: RECORDING RAW DATA

Qualitative: During the experiment I noted that most people exhibited the effects of caffeine within 5-10 minutes of consuming the 150ml black tea and coffee. As for when the people consumed the herbal tea, which possesses no caffeine, they seemed relaxed and calm.

Quantitative:

Table 2: Heart rate of the five subjects during rest, after consumption of Folgers Classic Roast with 60mg (mg ± 1.0) of caffeine, Caffeinated Twinings Black Tea with 34mg (mg ± 1.0) of caffeine, and Twinings Chamomile Tea with 0mg (mg ± 1.0) of caffeine.

Subjects	Trial	Resting (bpm± 1.0)	Caffeinated Coffee Folgers Classic Roast 60mg (mg ± 1.0) of caffeine in 150ml (ml± 1.0) bpm of subject (bpm± 1.0) 10 minutes after consumption	Caffeinated Twinings Black Tea 34mg of caffeine in 150ml (ml± 1.0) bpm of subject (bpm± 1.0) 10 minutes after consumption	Herbal Tea Twinings Chamomile Tea 0mg of caffeine in 150ml (ml± 1.0) bpm of subject (bpm± 1.0) 10 minutes after consumption
Subject 1	1	75.0	87.0	87.0	78.0
	2	78.0	102.0	90.0	81.0
Subject 2	1	84.0	90.0	87.0	84.0
	2	81.0	96.0	81.0	81.0
Subject 3	1	72.0	90.0	78.0	81.0
	2	75.0	99.0	90.0	78.0
Subject 4	1	60.0	72.0	87.0	63.0
	2	57.0	81.0	90.0	66.0
Subject 5	1	75.0	87.0	93.0	78.0
	2	78.0	84.0	87.0	81.0

Table 3: Average resting heart rate in bpm (bpm± 1.0), average heart rate after consumption of Caffeinated Folgers Classic Roast Coffee containing 60mg (mg ± 1.0) of caffeine. And the change in heart rate in bpm (bpm± 1.0). Range of change in heart rate is 17.0 bpm.

Subjects	Average resting heart rate (bpm± 1.0)	Average Caffeinated Folgers Classic Roast Coffee 60mg of caffeine in 150ml (ml± 1.0) bpm (bpm± 1.0) 10 minutes after consumption	Change in heart rate in bpm (bpm± 1.0)
Subject 1	76.5	94.5	18.0
Subject 2	82.5	93.0	10.5
Subject 3	73.5	94.5	21.0
Subject 4	50.5	76.5	26.0
Subject 5	76.5	85.5	9.0

Table 4: Average resting heart rate in bpm (bpm± 1.0). Average heart rate after consumption of Caffeinated Twinings Black Tea 34mg (mg ± 1.0) of caffeine. And the change in heart rate in bpm (bpm± 1.0). Range of change in heart rate is 36.5 bpm.

Subjects	Average resting heart rate (bpm± 1.0)	Caffeinated Twinings Black Tea 34mg of caffeine in 150ml (ml± 1.0) bpm (bpm± 1.0) 10 minutes after consumption	Change in heart rate in bpm (bpm± 1.0)
Subject 1	76.5	88.5	12.5
Subject 2	82.5	84.0	1.5
Subject 3	73.5	84.0	11.0
Subject 4	50.5	88.5	38.0
Subject 5	76.5	90.0	14.0

Table 5: Average resting heart rate in bpm (bpm± 1.0). Average heart rate after consumption of Twinings Herbal Chamomile Tea containin 0mg (mg ± 1.0) of caffeine. And the change in heart rate in bpm (bpm± 1.0). Range of change in heart rate is 11.0 bpm.

Subjects	Average resting heart rate (bpm± 1.0)	Twinings Herbal Chamomile Tea 0mg of caffeine in 150ml (ml± 1.0) bpm (bpm± 1.0) 10 minutes after consumption	Change in heart rate in bpm (bpm± 1.0)
Subject 1	76.5	79.5	3.0
Subject 2	82.5	82.5	0
Subject 3	73.5	79.5	6.0
Subject 4	50.5	64.5	14.0
Subject 5	76.5	79.5	3.0

Table 6: Table used to determine the t-test for independence.

	Amount of caffeine in grams, in beverages per 150ml serving.	Average change in heart rate in bpm (bpm± 1.0)
Folgers Classic Roast Coffee	60.0	16.9
Twinings Black Tea	34.0	15.4
Twinings Herbal Chamomile Tea	0.0	5.2

Standard deviation= .392127553
Null Hypothesis: The average change in heart rate in bpm (bpm± 1.0) is independent of the amount of caffeine consumed
Alternative Hypothesis: The average change in heart rate in bpm (bpm± 1.0) is not independent of the amount of caffeine

ASPECT 2: PROCESSING RAW DATA

Calculations:

Average of resting heart rate in bpm (bpm± 1.0): (Trial 1 + Trial 2)/2 = (75.0bpm + 78 bpm)/2 = 76.5 bpm

Range of change of heart rate in bpm (bpm± 1.0): For Coffee the highest change in heart rate in bpm was 26 bpm and the lowest was 9 bpm. Therefore the range was 26 bpm – 9 bpm = 17 bpm.

Standard Deviation of average resting heart rate: Use Excel, insert all values of resting heart rate in bpm, click an empty cell, type "=". go to function, type in "STDEV", highlight the values, click enter.

Standard Deviation
I conducted the standard deviation of heart rate in bpm of the individual beverages average change on heart rate to all be slightly different. For the Coffee I determine the standard deviation to be 7.83, for the black tea to be 2.8, and for the herbal tea to be 7.2. Therefore coffee had the greatest dispersion over the mean. This statistical calculation allowed me to obtain a more accurate estimation of the dispersion in the data. This is because outliers can skew data.

T-test

I conducted a t-test between the amount of caffeine in grams, in beverages per 150ml serving and the average change in heart rate in bpm (bpm± 1.0). This test allowed me to determine if the amount of caffeine statistically is independent or dependent of the amount of heart beats per minute produced after consuming caffeine. In order to calculate a t-test set up a table with independent and dependent values. Click an empty cell, go to function, type in "TTEST". high the values for the first array, and the for the second, and then depending on the project enter the number for tails, and type, then press enter.

Range

I chose to calculate range because I wanted to see the individual range of values per beverage. For coffee I determined there it to be 17.0 bpm, for black tea 36.5 bpm, and for herbal tea 11.0 bpm. The range shows how spread out the data are. Data with large ranges tend to be more spread out. This explains why the error bar for the average change in heart rate in bpm for the black tea is the longest.

ASPECT 3: PRESENTING PROCESSED DATA

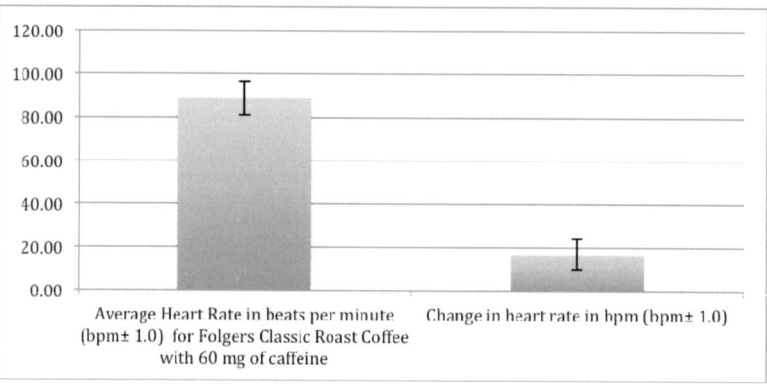

Figure 3. Average Heart Rate in beats per minute (bpm) for all five subjects ten minutes after consuming 150ml of Folgers Classic Roast Coffee containing 60mg of caffeine. Compared to the average change in heart rate in beats per minute (bpm).

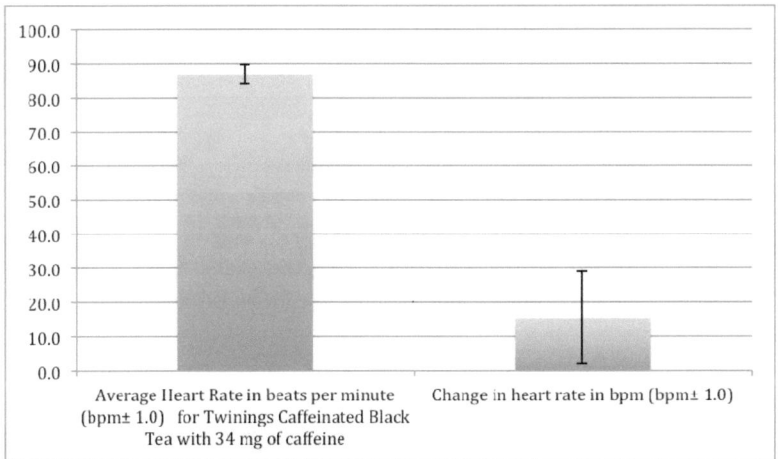

Figure 4. Average Heart Rate in beats per minute (bpm) for all five subjects ten minutes after consuming 150ml of Twinings Caffeinated Black Tea containg 34 mg of caffeine. Compared to the average change in heart rate in beats per minute (bpm).

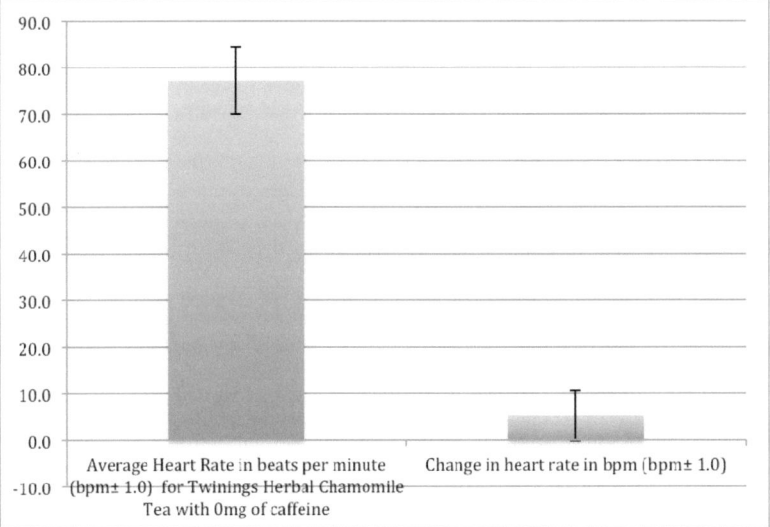

Figure 5. Average Heart Rate in beats per minute (bpm) for all five subjects ten minutes after consuming 150ml of Twinings Twinings Herbal tea containing 0mg of caffeine. Compared to the average change in heart rate in beats per minute (bpm).

Conclusion and Evaluation

ASPECT 1: CONCLUDING

In conclusion to my experiment, my hypothesis was proven correct. It was determined that if a human consumes a beverage containing a higher concentration of caffeine, then the heart rate will be higher than consuming a beverage with a lower concentration of caffeine. The average change in heart beat for the Folgers Caffeinated Coffee was 16.9 beats per minute (bpm) (bpm± 1.0), and contained the highest concentration of caffeine with 60 mg (mg± 1.0)of caffeine per 150ml (ml± 1.0), serving. The average change in heart beat for the Twinings Black Tea was 15.4 beats per minute (bpm) (bpm± 1.0), and contained the second highest concentration of caffeine with was 34mg (mg± 1.0) of caffeine for every 150ml (ml± 1.0) serving. And lastly the average change in heart rate for the Twinings herbal chamomile tea containing 0mg (mg± 1.0) of caffeine for every 150ml (ml± 1.0) serving was 5.2 (bpm) (bpm± 1.0). My range also explained how spread out the data was per individual beverage, and better explained why certain error bars were longer than others. Because of standard deviation I was able to conclude that coffee had the greatest dispersion over the mean. This statistical calculation allowed me to obtain a more accurate estimation of the dispersion in the data. This is because outliers can

I also used a T-Test on my data to reconfirm my hypothesis that the average change in heart beat, meaning the difference between the heart rate 10 minutes after consuming one of the three beverages minus the resting heart rate in bpm (bpm± 1.0), was not independent of the amount of caffeine per serving. In other words, I determined that caffeine had a direct relationship with the average change in heart beat, and it was not by coincidence, which supported my alternative hypothesis, which stated there was a statistical difference with scientific reason between the amount of caffeine ingested the average change in heart beat per minute (bpm). According to my t-test values, I am more than 95% confident that the manipulating (experimental) variable: the amount of caffeine ingested and the average change it has on heart beats per minute (bpm) is statistically different and significant through the experiment.

My results agree with published studies stating that caffeine concentration has a direct relationship with heart beats per minute (bpm). *In Tolerance to the Humoral and Hemodynamic Effects of Caffeine in Man* by David Robertson, et al. Heart rate increased an average of 6± 4 beats/min after meals associated with placebo and 7± 4 beats/min after meals associated with caffeine from tea, and double with caffeine from coffee.

This is because caffeine is a stimulant. Stimulants change the way the brain works by changing the way nerve cells communicate. Nerve cells, called neurons, send messages to each other by releasing chemicals called neurotransmitters. Neurotransmitters work by attaching to key sites on neurons called receptors. Stimulants can also cause the body's blood vessels to narrow, constricting the flow of blood, which forces the heart to work harder to pump blood through the body. The heart may work so hard that it temporarily loses its natural rhythm (*Stimulants*, 2012).

ASPECT 2: EVALUATING PROCEDURE

Systematic Error

The teabags could not be analyzed to reconfirm the exact amount of caffeine. Therefore one must trust that the value of the caffeine given on Twining's website is accurate. If the information is invalid there is no way of knowing, this can drastically impact the experiment in a negative way, because there would be too great of an uncertainty.

Human Error

The subjects may have drank or eaten a substance containing caffeine that the body was still trying to metabolize, and therefore had a higher resting heart rate.

The subject may have a tolerance against caffeine and therefore had a low change in heart beats per minute, possibly effecting the data.

Limitations

Some limitations of this experiment because it can be generalized for the "real world" and the majority of people. Meaning that there are many different types of beverages with individual caffeine content, that are not being accounted for in this investigation, such as energy drinks. All people metabolize caffeine at different rates and have a different response to caffeine concentration. And have moderately different heart beats per minute (bpm). People who have medical problems such as high blood pressure may display different results than someone who does not.

ASPECT 3: IMPROVING THE INVESTIGATION (parallel your sources of error and limitations with methods to improve. Also include further investigation possibilities)

There are a few minor adjustments that could be made in this investigation which would further change the procedure and generate more accurate results.

Before a subject takes part in this experiment they must fill out a questionnaire. This questionnaire will ask if the person is a chronic drinker, has a predisposed medical condition etc. This will better help determine which subjects are deemed "most fit" to participate in this experiment. And also allow me to make sure all groups of people are being represented. For example male, female, active, sedentary, etc. Therefore one does not create a conclusion based on a small sample size. One can make a more informed conclusion that can better represent the people of the world.

REFERENCES AND CITATION

Adan, A. (2008). Early effects of caffeinated and decaffeinated coffee on subjective state and gender differences. *Progress in Neuro-Psychopharmacology and Biological Psychiatry, 32(7), 1698-1703*

What's a normal resting heart rate? (2012, September). Retrieved from http://www.mayoclinic.com/health/heart-rate/AN01906

Medline Plus: Caffeine. (2013, January). Retrieved from http://www.nlm.nih.gov/medlineplus/caffeine.html

Robertson, D (1981), Tolerance to the Humoral and HemodynamicEffects of Caffeine in Man. *The American Society for Clinical Investigation, Inc. 67(4), 1111-1117*

Stimunalnts (2012, March), Retrieved from http://teens.drugabuse.gov/facts/facts_stim2.php

Starr, C (2009), Biology: Today and Tomorrow With Physiology, Retrieved from http://books.google.com/books?id=dxC27ndpwe8C&printsec=frontcover#v=onepage&q